# MERCURY

## SEYMOUR SIMON

MORROW JUNIOR BOOKS
New York

PHOTO AND ART CREDITS
Photograph on page 4–5 courtesy Dewey Vanderhoff; all other photographs
courtesy Jet Propulsion Laboratory/National Aeronautics and Space Administration.
All artwork by Ann Neumann.

The text type is 18 point ITC Garamond Book.

Printed in Singapore at Tien Wah Press.
3   4   5   6   7   8   9   10
Library of Congress Cataloging-in-Publication Data
Simon, Seymour
Mercury / Seymour Simon.
p.      cm.
Summary: Describes what is known about Mercury from the
photographs taken by Project Mariner.
ISBN 0-688-10544-0 (trade)—ISBN 0-688-10545-9 (lib. bdg.)—0-688-16382 (pbk.)
1. Mercury (Planet)—Juvenile literature.   2. Project Mariner—
Juvenile literature.   [1. Mercury (Planet)   2. Project Mariner.]
I. Title.
QB611.S56   1992
523.4'1—dc20   91-17404   CIP   AC

For Michael and Deborah

Mercury is hard to see from Earth, even though it sometimes looks brighter than any of the stars in the sky. But Mercury is a planet, not a star. It appears bright because it comes closer to us than does any other planet except Venus and Mars. Because it orbits so close to the sun, we do not see Mercury against the dark background of the night sky. Mercury is visible only during twilight hours, either low in the east just before sunrise or low in the west just after sunset.

The early Romans named Mercury after the messenger of their gods because it appeared to move more quickly through the sky than did any of the other planets.

Mercury is the planet closest to the sun. Its average distance from the sun is 36 million miles, about one-third of Earth's 93 million miles. Mercury travels around the sun more quickly than any other planet. Earth orbits the sun in 365 days, or one Earth year. Mercury takes 88 Earth days to orbit the sun, or one Mercury year.

Mercury is the second-smallest planet (after Pluto), just 3,030 miles across. If Earth were hollow, eighteen planets the size of Mercury could fit inside. In fact, Mercury is smaller than Jupiter's and Saturn's largest moons; Mercury has no moons of its own.

Mercury, Venus, Earth, and Mars are called the inner planets. These four rocky planets are much smaller than the four giant outer planets—Jupiter, Saturn, Uranus, and Neptune—which are made up mostly of gases.

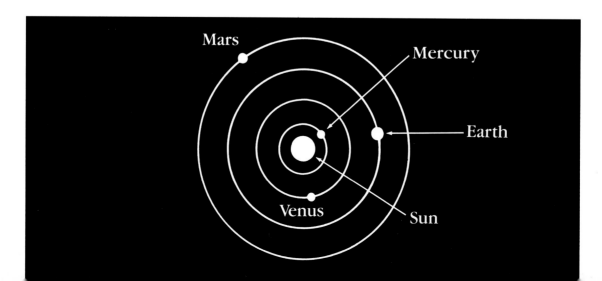

MARS

MERCURY

MOON

EARTH

IO

EUROPA

GANYMEDE

CALLISTO

VENUS

TITAN

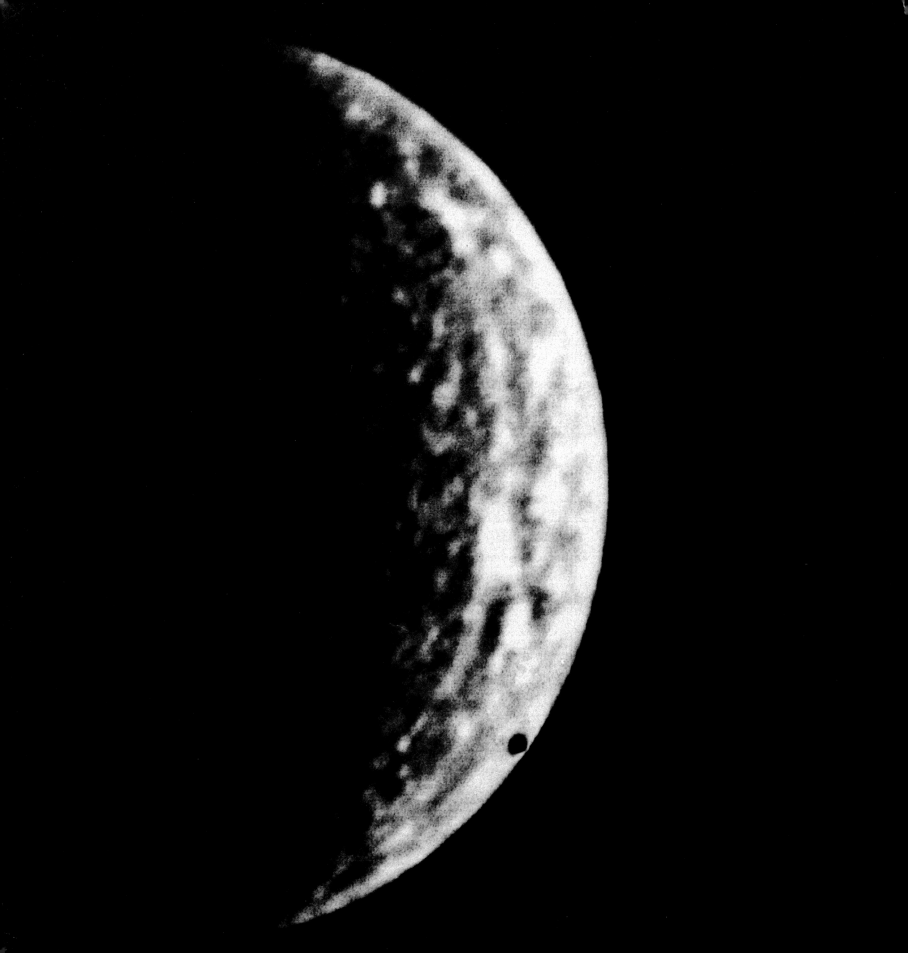

Viewed from Earth, Mercury appears to change its shape from day to day, much the way our moon and the planet Venus do. When it is close to us, Mercury looks about three times bigger than it does when it is on the opposite side of the sun from us. But Mercury looks small even when seen through a telescope, and it is difficult to photograph from Earth.

Not much was known about Mercury until it was studied by radar from Earth in the 1960s and, later, visited by a space probe, *Mariner 10*, in March and September of 1974 and in March of 1975. *Mariner* found that Mercury takes 59 Earth days to spin once on its axis, or one Mercury day.

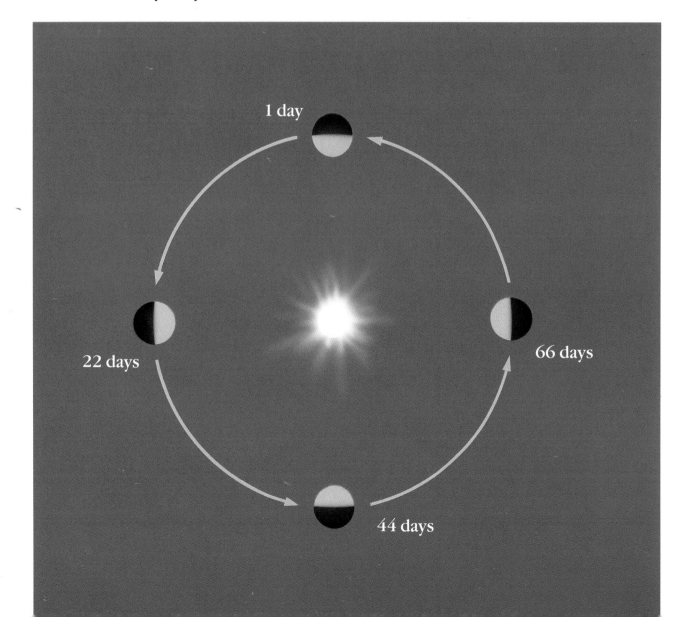

1 day

22 days

66 days

44 days

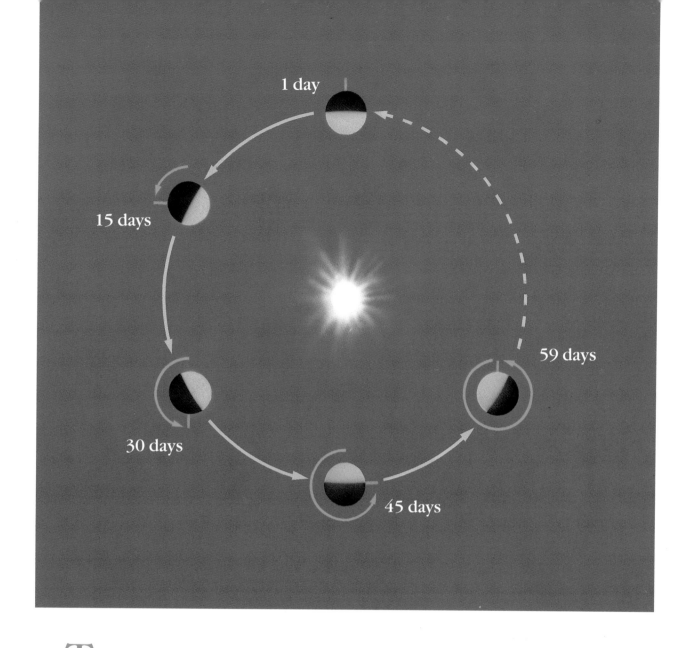

1 day

15 days

59 days

30 days

45 days

That's almost as long as one Mercury year—the 88 days it takes to orbit around the sun. Because Mercury is moving around the sun in the same direction as it is slowly spinning, both daylight and nighttime on Mercury last a very long time.

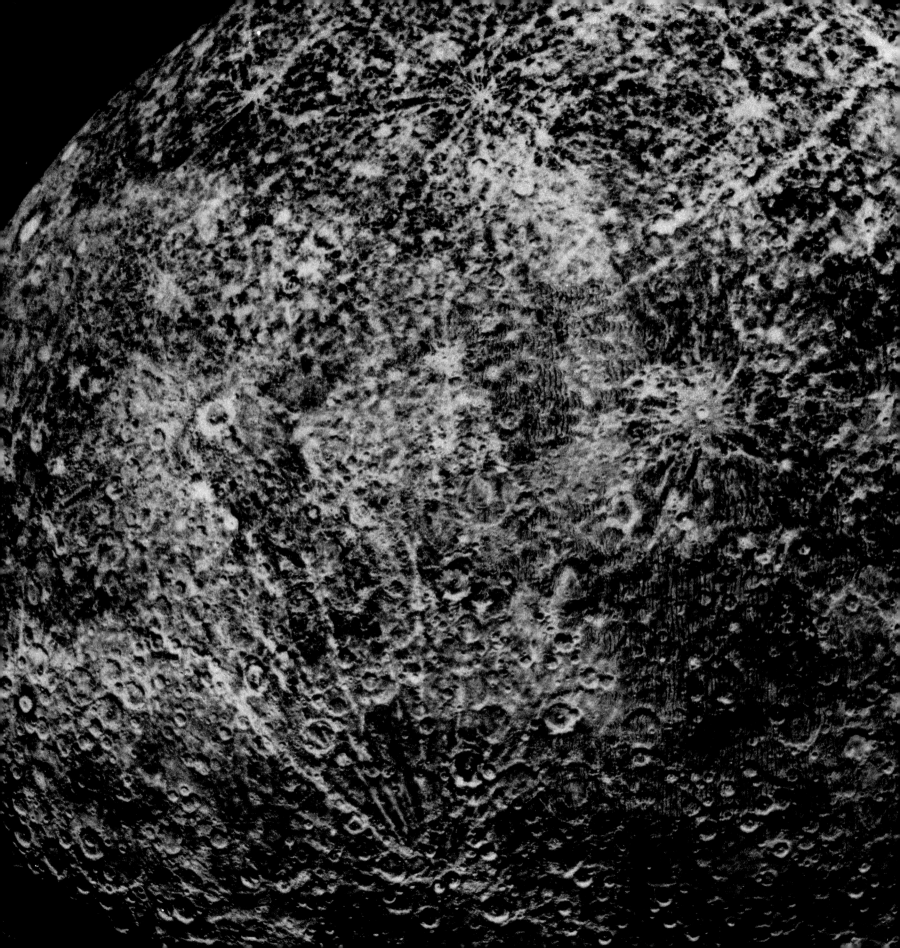

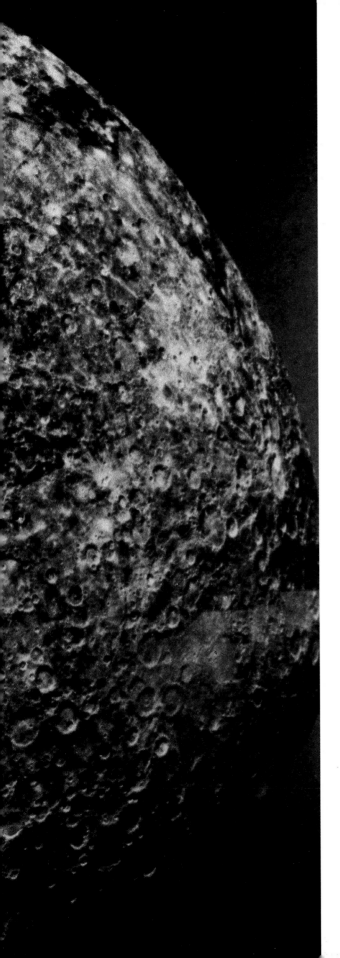

Mercury is a small planet, but it is a heavy one for its size. Mercury's density is 5.4, which means the planet is 5.4 times heavier than an equal amount of water. That's nearly the same as Earth's density of 5.5 and greater than the 3.9 density of Mars.

The reason for Mercury's density is its enormous, heavy core of molten iron, about 2,250 miles across. Above this core is a layer of squeezed molten rock— similar to the lava that erupts from volcanoes on Earth. A solid rocky surface, or crust, floats on this underlying layer. In our Solar System, only Mercury, Venus, Earth, and Mars (the inner planets) have rocky surfaces.

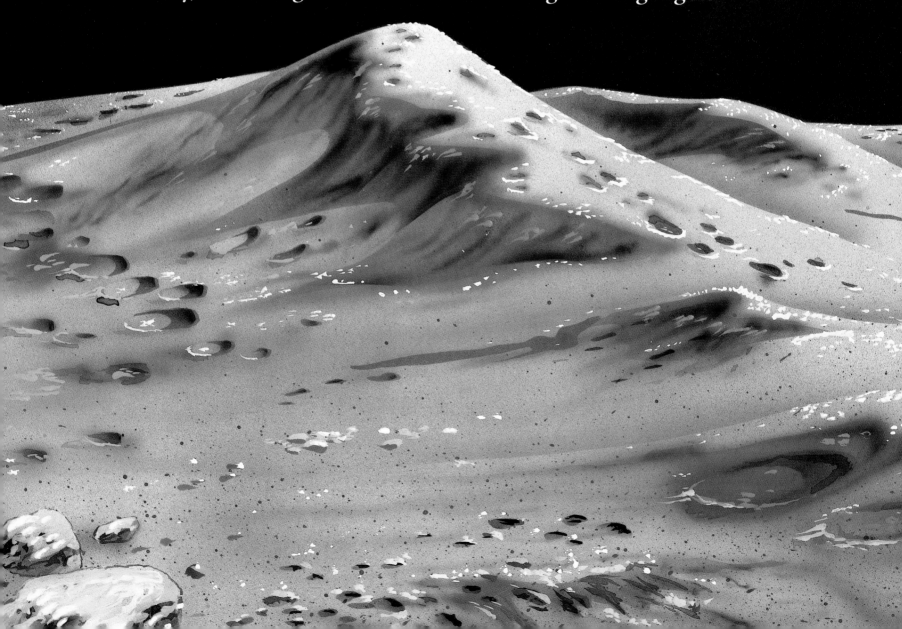

Unlike Earth and Venus, Mercury is an almost airless planet. On Earth and Venus, the atmosphere acts as a blanket, so surface temperatures do not change greatly from day to night. Because Mercury is very near the sun, the temperature rises to over 750 degrees (F) during the day, hot enough to melt lead. Yet during the long nights (roughly three Earth months long), with no atmosphere to trap the heat, the temperature on Mercury drops to − 300 degrees (F), colder than the coldest temperature any place on Earth's surface. Mercury's day-to-night temperature change of over 1000 degrees (F) is greater than the temperature change on any other planet.

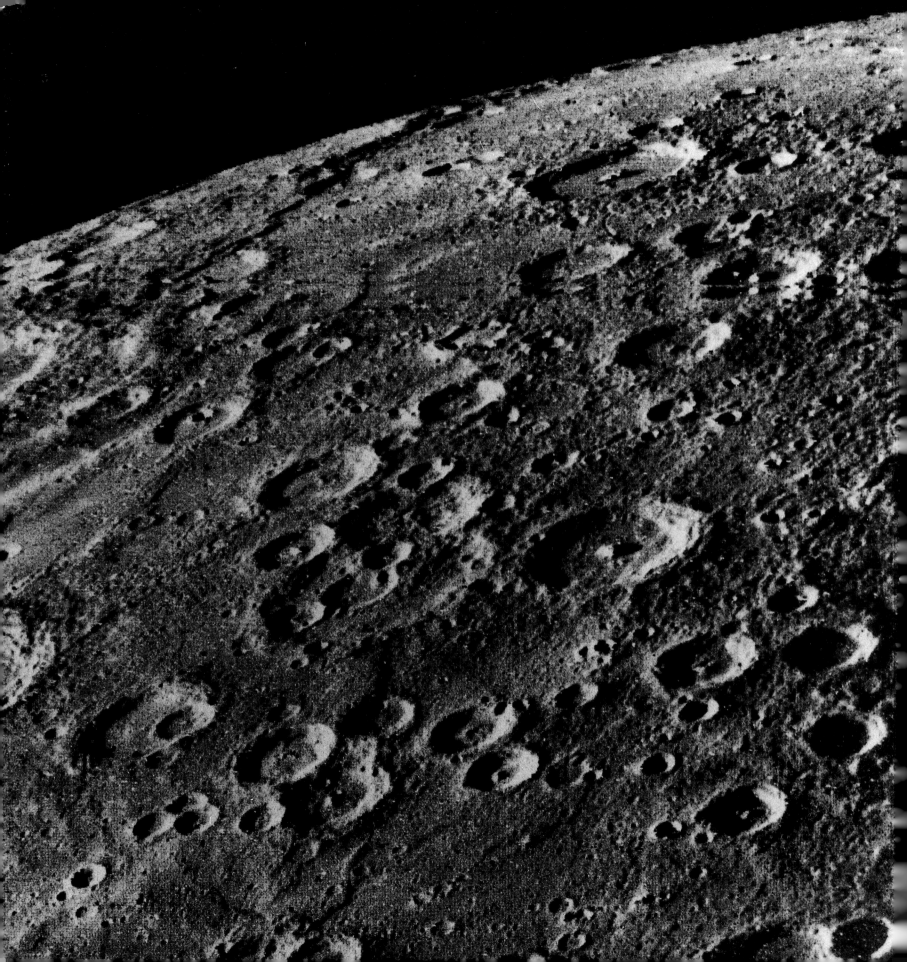

The surface of Mercury looks very much like the heavily cratered surface of our moon. The large craters were made during the past 4 billion years by countless meteorites or asteroids crashing into the planet's surface, which is not protected by an atmosphere. The largest impact craters are more than a hundred miles wide. Mercury has no large areas of highlands or lowlands as do Earth and Venus.

Thousands of smaller, bowl-shaped craters speckle the landscape. Many of these were made when rocks thrown up from the impact of the meteorites and asteroids came crashing back to the surface of the planet.

As *Mariner* approached Mercury at a speed of nearly seven miles per second, it took this photo from only 21,000 miles away. It shows a heavily cratered surface with many low hills. The large valley on the left is more than four miles wide and sixty miles long. The crater on the right is about fifty miles across.

Because Mercury is denser than our moon, its craters are shallower. The large craters on Mercury seem at some time to have been flooded with lava, which hardened to form smooth, flat basins.

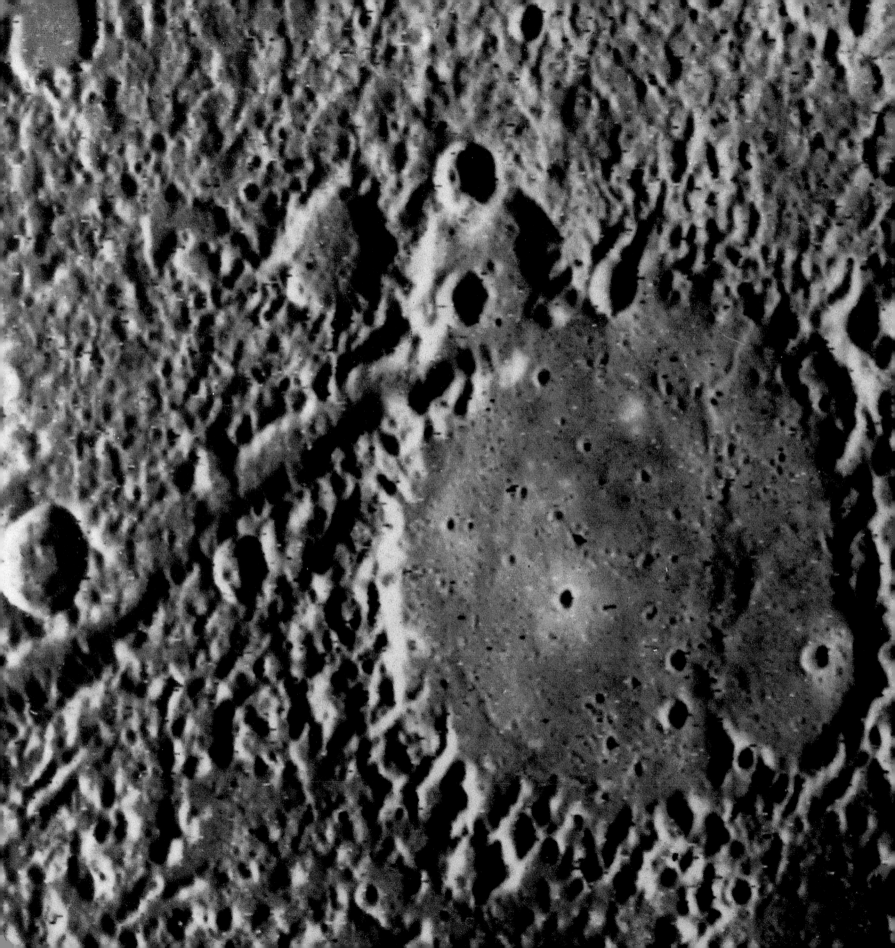

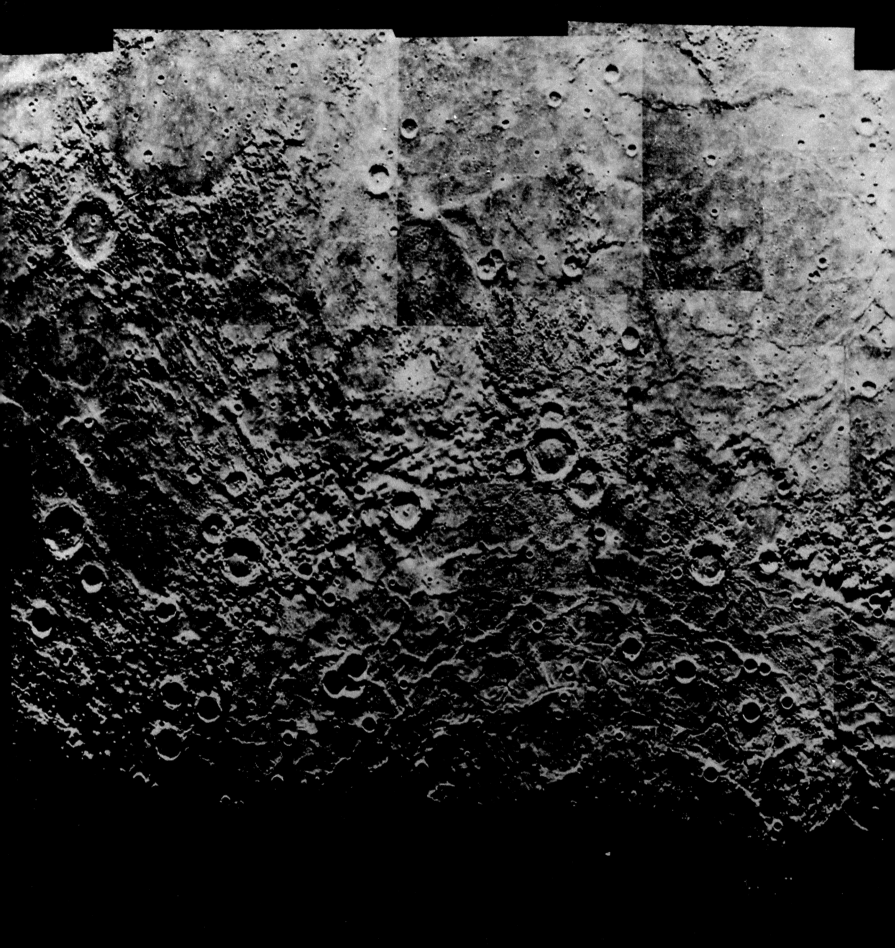

The largest surface feature on Mercury is the Caloris Basin, which can be seen in the lower part of this image. A circular plain about eight hundred miles across, the basin is ringed by mile-high mountains. Its floor is heavily cracked and covered by many hills or ridges. Caloris was probably carved out of the surface of Mercury by the impact of a large meteor or asteroid many millions of years ago. Over the years, lava flooded its interior. The smaller craters were made after the basin was formed. This image was made by piecing together eighteen *Mariner* photos.

Many long lines of high cliffs, called scarps, cut across the surface of Mercury. The scarp in this photo is nearly two hundred miles long. It slashes through craters and whatever else lies in its path. The scarps may have been caused when the interior of the planet cooled and the overlying crust buckled and cracked.